COLD CLIMATES

Keith Lye

THE WORLD'S CLIMATES

COLD CLIMATES
DRY CLIMATES
EQUATORIAL CLIMATES
TEMPERATE CLIMATES

Produced for Wayland (Publishers) Limited
by *specialist* publishing services, London

Series designer: Eric Drewery

First published in 1996 by Wayland (Publishers) Limited,
61 Western Road, Hove
East Sussex BN3 1JD England

British Library Cataloguing in Publication Data
Lye. Keith, 1933 -
Cold Climates. - (The World's Climates)
1. Human ecology - Cold regions - Juvenile literature
2. Ecology - Cold regions - Juvenile literature 3. Cold
regions - Climate - Juvenile literature 4. Polar regions -
Climate - Juvenile literature
I. Title
574.5'2621

ISBN 0 7502 1816 9

Printed and bound by G. Canale, Turin, Italy.

Picture acknowledgements
B&C Alexander: pp 8/9, 13, 14, 15, 18, 24,
28, 29, 34, 35, 42; Frank Lane Picture
Agency: p 35; Heather Angel: p 33;
Liz Draper: pp 1, 37, 39; Oxford
Scientific Films: pp 11/12, 13, 16, 21, 23,
25, 30, 32, 36; Tony Stone Images: p 27;
Wayland: p 5.

Contents

Lands of snow and ice

COLD REGIONS

When we talk about climate we mean the usual, or average, weather of a place. If we know about the climate of a place we want to visit, we will pack the right clothes when we go there. Climate determines what plants and animals are found in an area. It also influences how people live.

One major factor affecting the climate of a place is how much heat it gets from the sun. The sun's heat is strongest at the equator, in the central part of our world. Near the North and South Poles, the sun's rays are spread over a much larger land area. They also have to pass through a thicker layer of air which absorbs a large amount of the sun's heat. As a result, less heat reaches the ground. This makes places near the equator warm, while areas around the poles are bitterly cold.

The world has four main climates. They are cold climates, equatorial (hot and wet) climates, dry climates and temperate climates. Temperate climates are neither too hot, too cold nor too dry.

This book is about cold climates. The world has four main cold regions. One area includes all the ice and snow-covered lands around the North and South Poles. The second cold region is a treeless area called the tundra, which has a short summer season when plants grow. A third is called the subarctic, or cold temperate region. This area is known for its vast forests of hardy cone-bearing trees.

Highlands form the fourth of the world's cold climate regions. Even in hot countries, high mountain tops are cold places, capped by snow and ice.

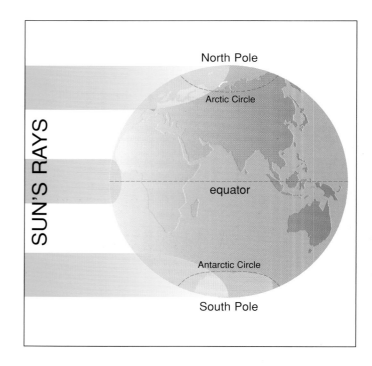

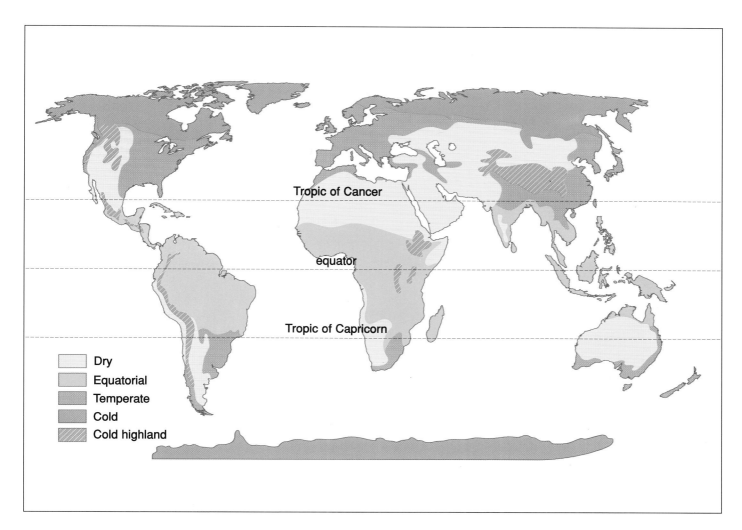

Tropic of Cancer

equator

Tropic of Capricorn

- Dry
- Equatorial
- Temperate
- Cold
- Cold highland

Above The main climate regions of the world. Cold climates also occur locally at very high altitudes.

Right Ice at the top of Mount Kenya, which is situated near the equator.

Left The effect of the sun's rays at the poles and at the equator, where it is more direct and hotter.

LANDS OF THE MIDNIGHT SUN

Our lives are ruled by the sun – during the day we work and play, and by night we sleep. At the North and South Poles, the sun does not set for half of the year. For the other half, it does not rise.

If you look at a globe of the world, you will see a network of imaginary lines. The line that runs around the middle of the globe is the equator. It divides the world into two equal halves called hemispheres and is numbered 0 degrees latitude. Other lines of latitude run around the globe between the equator and the North Pole (at 90 degrees north of the equator) and the South Pole (at 90 degrees south).

One line of latitude near the North Pole is called the Arctic Circle. It runs around the globe at $66\frac{1}{2}$ degrees north of the equator. The Antarctic Circle is $66\frac{1}{2}$ degrees south of the equator. At places on these lines, the sun does not set for one or two days every year.

The region between the Arctic Circle and the North Pole is called 'the land of the midnight sun'. A similar region lies south of the Antarctic Circle. Every place in

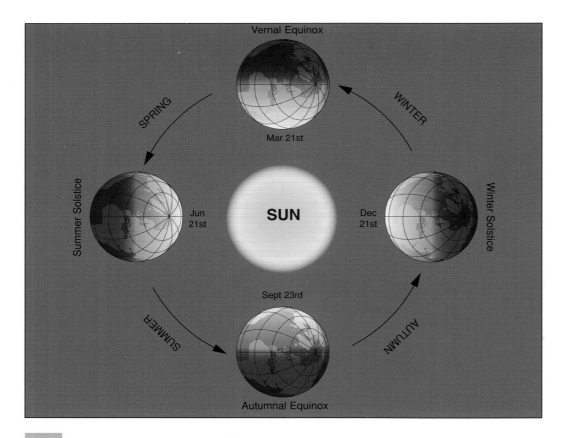

Left *How the tilt of the earth affects sunlight during the year. During December, the northernmost part of the earth is constantly facing away from the sun. By June the situation is reversed, so that in some northern places during midsummer the sun never sets.*

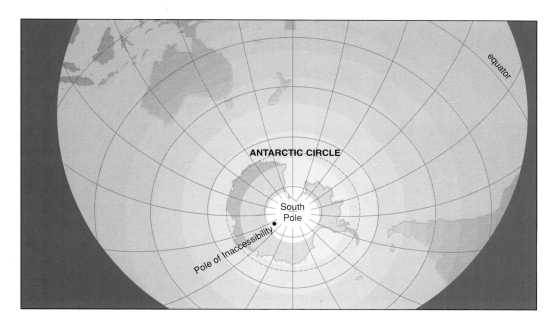

Left The Arctic Circle.

Left The Antarctic Circle.

these regions has between one and 183 days every year when the sun does not set.

The midnight sun is caused by a tilt in the earth's axis of $23\frac{1}{2}$ degrees. The axis is the imaginary line through the earth that joins the North Pole, the centre of the earth and the South Pole. As our planet rotates, first the northern and then the southern hemisphere leans towards the sun. When the North Pole is slanted towards the sun, the lands around the Arctic Circle enjoy a brief summer. At this time the South Pole leans away from the sun and Antarctica has a dark, cold winter.

Polar lands

ICE SHEETS AND ICE CAPS

The ice-covered regions near the poles are the coldest places on earth. The South Pole lies in the middle of Antarctica, a vast continent which is bigger than either Europe or Australia. A huge ice sheet covers about 98 per cent of Antarctica. In places, the ice is 4800 m thick, which is about 16 times the height of the Eiffel Tower in Paris.

Below *Aurora borealis (Northern Lights) at Skarsvag, Norway, showing the rarely seen red band.*

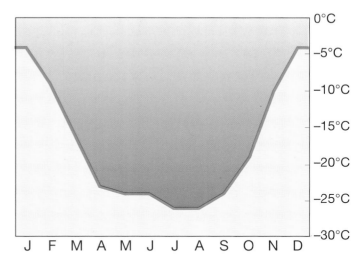

Left *Sea ice in the Arctic Ocean.*

Below *Average temperature chart for McMurdo Sound in Antarctica.*

A Russian research centre in Antarctica, called Vostok Station, holds a world record. In 1983 the world's lowest air temperature was recorded here: −89.2 °C. But the place with the coldest climate is the Pole of Inaccessibility, measured by its average yearly temperature of −58 °C. This is a point, near to the South Pole, which is farthest from all of Antarctica's coasts.

The North Pole lies in the middle of a frozen ocean. Here temperatures are not as low as in Antarctica. However, several bitterly cold land areas in Europe, Asia and North America, including many islands, lie in and around the Arctic Ocean. The world's second largest ice sheet covers about four-fifths of Greenland, the world's largest island. Average temperatures in central Greenland range from about −47 °C in February to −11 °C in July. The Arctic region also contains many smaller bodies of ice, called ice caps or glaciers.

Polar regions are known for the striking displays of lights that sometimes appear in the sky at night. In the Arctic region they are called the Northern Lights, or the *aurora borealis,* and in the southern hemisphere they are called the Southern Lights, or the *aurora australis.* These beautiful lights are caused when streams of particles from the sun collide with atoms and molecules in the upper atmosphere, high above the polar regions.

SNOW AND WIND

In icy polar regions, average temperatures in the warmest month never rise above 0 °C (freezing point). Although the regions around the North and South Poles have up to six months of continuous sunlight every year, the sun is low in the sky and its heat is spread over a large area. Also, much of the solar energy that reaches the surface is reflected back into space by the snow and ice. On average, Antarctica is three times higher than any other continent. This contributes to its extreme climate.

Above *Winds cause snow drifts in Antarctica.*

Polar regions are windy places. The air above them is cold and dense. The air sinks downwards, creating large areas of high air pressure at the surface. From these areas, winds blow outwards. These winds are often strong and they make it feel much colder than it is. This is called the wind chill factor. Strong winds often blow loose snow and tiny crystals of ice across the surface, creating storms called blizzards.

Above *During a blizzard in Antarctica it is difficult to see more than a few metres ahead.*

Left *A view of some of the many mountains in Antarctica which contribute to making it on average the highest continent on earth.*

Blizzards are one of the most unpleasant features of icy polar climates. While they last, people cannot see to travel.

Another feature of the icy polar climates is the low average yearly precipitation. Precipitation is the name for all forms of water which come from the air, including rain, snow, sleet, hail and frost. In polar regions, precipitation occurs most commonly as snow. The total precipitation in icy polar regions is generally less than the equivalent of 250 mm of rain. As a result, polar regions are cold deserts. They have about the same yearly precipitation as hot deserts.

Most of the snow does not melt. Instead it piles up, layer upon layer, until it is compacted (pressed) into ice. Gravity causes the ice to move outwards towards the coasts. There, chunks of ice break away to form floating islands called icebergs.

PLANTS AND ANIMALS

Plants are rare in places with icy polar climates. Scientists have found only two kinds of flowering plants in Antarctica. One is a grass and the other is a herb. Both of these plants grow only on ice-free areas in the Antarctic Peninsula, which extends towards South America.

However, about 340 simple plants, such as lichens, mosses, toadstools and liverworts, have been found in Antarctica. They live mainly on sheltered, north-facing slopes, where they can catch the sun. There are about 200 lichens. Lichens are organisms which consist of an alga and a fungus living together as a single unit. They have no roots and they cling to rock surfaces. Some lichens, algae and fungi grow inside minute cracks in the rocks.

Only a few tiny creatures, such as flies and midges, spend all their life in Antarctica. In contrast, the seas around the continent are rich in fish, krill (small shrimp-like animals), dolphins and whales. Seals and penguins, which nest near the coasts, as well as other animals, feed on the fish and krill. Although they are flightless and awkward on land, penguins are magnificent swimmers. The largest of these, the emperor penguin, nests inland far from the sea. During the 100 or so days it takes to rear and hatch its chick, it does not eat and loses up to 40 per cent of its body weight.

The best known animal which lives on the floating ice around the Arctic Ocean is the polar bear. This animal is well adapted to survive in

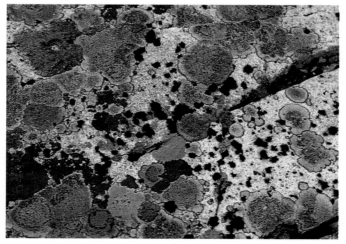 

Above *Lichens (left) and mosses (right) found in Antarctica.*

Above Gentoo penguins in Antarctica.

Left Polar bears are kept warm by their thick coats and extensive layers of fat.

the cold climate. Its thick white coat and layers of body fat help it to keep warm. It is also well camouflaged when it goes hunting seals and other animals. In winter, polar bears often live in dens dug out of the snow. However, most Arctic animals live south of the icy regions, in the tundra and subarctic regions.

PEOPLE IN POLAR REGIONS

People called Inuit live in the frozen wastelands of northern Canada and Greenland. They are closely related to other similar people – the Inupiat and Yupuk in Alaska, and the Yuit in Siberia. Together, these people number about 120,000.

Until the arrival of Europeans, the life of the Inuit and the other northern people was well adapted to the icy climate. They wore warm clothes made from animal skins

Above *Fishing for polar cod through an ice hole in Canada.*

and they built snow houses as shelters against the winter cold. They lived by hunting animals such as polar bears and seals around the Arctic Ocean, and caribou in the tundra to the south. They cut holes in the frozen ocean through which they speared fish. On land they travelled on sledges drawn by dogs, and on the water they used kayaks (boats with wooden frames covered by skin). European explorers paid Inuit people to travel with them, learning how to survive in the icy northern wastelands.

The traditional Inuit way of life survived until the 1950s. Then the Canadian government encouraged the people to move into more permanent homes. Today most Inuit live in towns, working in the fishing or construction industries or in other jobs.

Unlike the Arctic, Antarctica has no permanent population. Scientists from about ten nations work there doing research into the geology and climate of the continent. The largest scientific research station is the American McMurdo Station. In summer, the station houses about 1,000 people, though only about 200 remain during the dark winter months.

Left An Inuit police officer on patrol near Baker Lake, Canada.

Below A scientist making weather observations using a meteorological balloon at Halley Station, Antarctica.

The scientists in Antarctica have to get all their supplies, including food and fuels, shipped or flown in from the outside world. This enables them to live comfortably even when fierce blizzards are raging outside. They have shown that, with modern technology, people can live anywhere, even in the harsh climate of the South Pole.

The tundra

PERMAFROST

From time to time, scientists working in the tundra regions of Siberia dig up the complete bodies of woolly mammoths. These prehistoric animals, which were like hairy elephants, died out 10,000 years ago. The reason why their bodies have survived is that they were buried in subsoil, which is always frozen down to depths of 600 m. The frozen subsoil is called the permafrost. It acts like a natural refrigerator.

Tundra climates have bitterly cold winters and short chilly summers, when fogs are common. The average temperature in the warmest month rises above freezing point, though not above 10 °C. The warmer temperatures occur for only two to four months every year. During this brief period

Above *Permafrost in a tunnel – the sand layers have been shaped by the ice.*

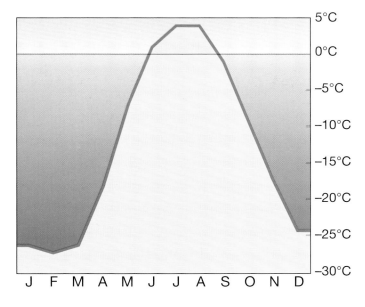

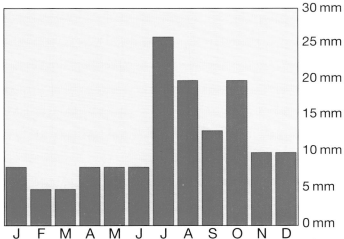

Temperature and rainfall graphs for Barrow Point, Alaska with little precipitation and very low temperatures.

the top layer of the soil, up to 80 cm deep, thaws out, while the permafrost below remains frozen hard.

The precipitation, as in the icy polar regions, is low. Over most of the tundra, the average annual precipitation is under 300 mm. The air is so dry it can make people become extremely thirsty.

Some rain or wet snow falls in summer, while powdery snow coats parts of the land in winter. Strong winds create blinding blizzards in the tundra. The winds often sweep away almost all the snow from some areas. In other areas the snow piles up in drifting heaps.

Tundra occurs only in the northern hemisphere. This is because oceans occupy the parts in the southern hemisphere where we would expect to find tundra. The most severe tundra climates occur in north-eastern Siberia, especially in remote areas far from any coast. Here average temperatures in the coldest month may fall to below −40 °C. The world's coldest village, Oymyakon, is in Siberia. The 4,000 people who live there have to put up with winter temperatures that sometimes reach −70 °C.

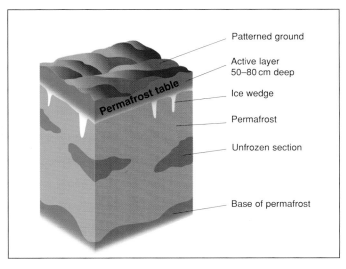

Above *Permafrost is frozen subsoil.*

PLANTS AND ANIMALS

The northern edge of the tundra marks the end of most plant life. The southern limit is another plant boundary, marking the northern limit of tree growth. In between lies the treeless tundra region.

During the short summer the frozen tundra surface melts. It becomes swampy because the permafrost below stops water draining down through the ground. The word tundra means 'barren land' in Finnish, but in summer the land is covered by a carpet of plants. These include lichens, mosses, sedges (grass-like plants), low shrubs, occasional dwarf trees, and many flowering plants. All the plants are small and some are evergreens. They hug the ground to avoid damage by strong winds. Tundra plants must grow quickly. They start to grow as soon as temperatures rise above freezing point. Some grow from shoots and flower buds left over from the previous year.

During the short growing season swarms of black flies, midges and mosquitoes fill the air. The insects

Left Tundra landscape of North-west Greenland during the summer with some melt streams.

Above An arctic tern hovering before diving for fish.

18

Left Some animals change colour to blend in with their surroundings. When the winter snow starts to fall, these Arctic hares' coats change from brown to white.

Left A musk ox on Wrangel Island, Russia. Musk oxen get their name from the musky smell of their coat. They have a long, dark-brown outer coat over a soft, dense, light-brown undercoat.

pester and bite people, but they are also food for many birds, such as geese and terns, which arrive from the south to nest in the tundra. The Arctic tern is a famous traveller. Some fly more than 35,000 km every year, from the Antarctic Circle to the Arctic Circle and back again.

Grazing animals, such as caribou and reindeer, also arrive. They come from the forests in the south to feed on the tundra plants. Other animals include Arctic foxes and hares, grizzly and polar bears, lemmings, musk oxen, wolves and, on the coast, seals and walruses.

When autumn comes, caribou, reindeer and most birds move south. Other mammals, like the Arctic hare, grow white winter coats for camouflage. Some birds of prey, such as gyrfalcons and snowy owls, also winter in the tundra.

PEOPLE OF THE TUNDRA

The people of the tundra used to live mainly by hunting and fishing. In North America, the main people of the tundra were groups of Inuit, who moved around hunting caribou and other animals. Like other Inuit to the north, most of the people in the Canadian tundra are now settled in towns.

The Sami (or Lapps) of northern Norway, Sweden, Finland and north-western Russia were divided into three main groups. One group lived by fishing on the coasts, while another lived in the forests south of the tundra. A third group traditionally reared reindeer for a living. They followed the herds from the northern forests to the tundra in summer. They relied on them for almost all their needs. These animals were their main source of meat and milk, while their skins were used for clothing and boots. Reindeer also carried heavy loads and pulled sledges. Today only a few Sami follow their ancient wandering way of life.

Left Nentsey people with their reindeer in northern Siberia.

Right Inuit women and children in Canada.

Most of them now earn their living in a variety of jobs in the towns.

Other groups of people, such as some Chukchi and Yakut, live in the tundra regions of north-eastern Siberia. They, too, once led wandering lives, relying on hunting, fishing, fur trapping and rearing reindeer herds. But from the 1920s, the Communist government of the former Soviet Union persuaded the people to settle in villages or large farms.

Iceland is a country just south of the Arctic Circle. The higher parts of the island contain ice-caps surrounded by tundra, but the coasts have a surprisingly mild climate. This is because a warm ocean current, called the Gulf Stream, passes along the shore, warming onshore winds. The people who live along the coast have developed a modern fishing industry and are among Europe's most prosperous people.

Cold temperate climates

SUBARCTIC LANDS

If you fly directly from London to Los Angeles you can view three cold climate zones through the aircraft's windows. First you fly over Iceland and then across the huge Greenland ice sheet. Then you reach the flat, treeless tundra of northern Canada. As your aircraft turns south, you will spot isolated trees growing on sheltered, south-facing slopes. Gradually the number of trees increases until you are over the northern forests of Canada's subarctic region.

The subarctic region has a cold, temperate climate. The average temperature in the coldest month is always below −3 °C. The average temperature in the warmest month is above 10 °C. Generally, winters are long and severe, especially in inland areas far from the sea. Lakes and seas freeze over. For example, the Baltic Sea between Sweden and Finland is closed in winter to all ships other than ice-breakers. The same is true of the gulf of the St Lawrence river in Canada. There are short spring and autumn seasons, while summers are longer and warmer than in the tundra. The precipitation is generally light.

This region occurs only in the northern hemisphere. This is because oceans fill the areas in the southern hemisphere where subarctic regions would occur.

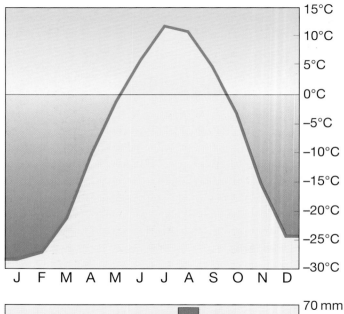

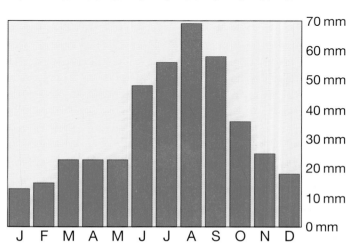

Churchill, Canada has longer, warmer summers than the tundra.

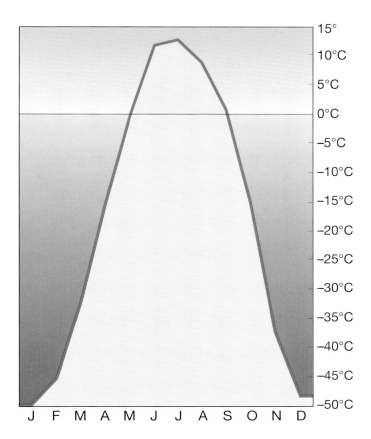

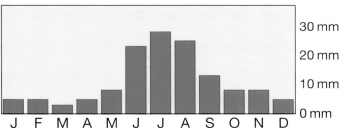

Verkhoyansk, Siberia has an extreme subarctic climate.

Above *An ice breaker in the north Baltic Sea.*

A broad subarctic region stretches across Canada, reaching as far south as Lake Winnipeg and the northern shores of Lake Superior. In Europe, the cold temperate climate includes northern Scandinavia and northern Russia. In Asia, the area stretches from the tundra in the north to northern Mongolia and north-eastern China.

Siberia has some of the most extreme subarctic climates. The difference between average temperatures in the coldest and warmest months may be more than 60 °C. Frosts are common and may occur even in July and August. As a result, farming is not possible in most areas and the subarctic is thinly populated.

THE TAIGA

The subarctic (or cold temperate) climate zone is a region of huge forests. Their Russian name is the *taiga*, and they are also known as the boreal (northern) forests. The chief plants in these forests are needleleaf evergreen trees, which are also called conifers. The main trees are fir, pine and spruce, though larches (deciduous trees which shed their leaves in winter) are also common, especially in eastern Siberia. Mixed forests of both conifers and deciduous trees occur in the warmer parts of the southern taiga.

The boreal forest stretches across Canada, and also from the Baltic Sea to the Pacific Ocean. The Eurasian taiga is the world's largest forest. It is more than three times bigger than the rain forest in the Amazon basin in South America. In both Canada and in the Eurasian taiga there are swampy areas and broad, muddy river valleys.

The needleleaf trees reach heights of about 20 m. Where sunlight reaches the ground, shrubs grow, including many small bushes which bear berries. Mosses and lichens grow on the forest

Left *Reindeer amongst stunted trees of the Eurasian taiga.*

Right *A coniferous forest of fir, pine and spruce in northern Russia.*

floor and also on the tree trunks. There are also a few flowering plants, but large areas of the forest floor are blanketed by needles. Because of the harsh climate, these forests contain far fewer species (kinds) of plants than the forests in the hot and wet equatorial regions.

The trees in the taiga are adapted to survive the bitterly cold, snowy climate. Most have shallow roots that obtain moisture from the topsoil, even when the subsoil is frozen. Their thick barks protect them against the cold, while their conical shapes prevent overloading by snow. Apart from larches, the evergreen trees can resume their growth as soon as it is warm enough. In such ways, they make the best use of a short growing season.

ANIMALS OF THE TAIGA

The taiga is surprisingly rich in animal life. This is mainly because there are lakes and marshy areas as well as forest. These offer varied conditions for animals. The taiga has enough food resources to enable most animals to spend the entire year there. Only a few, such as the caribou, move to the tundra in the summer. Small mammals include chipmunks, mice, squirrels and voles. Many are adapted to live under the snow in winter.

Several species of birds live in the taiga. Some, such as crossbills, nutcrackers and woodpeckers, are skilled at extracting seeds from the woody cones of the coniferous trees. Seeds are a major food in the forests. Some animals, such as chipmunks and squirrels, hoard stocks of cones so that they have a supply of food in winter. Birds also feed on the insects that swarm through the taiga in summer. In winter, they feed on dormant

Left *A North American timber wolf blending into its snowy surroundings.*

insects hidden in cracks in the barks of trees. The most common birds in the forests are members of the grouse family. They feed on berries and buds in summer. In winter they eat pine needles.

Large mammals include black and brown bears, caribou, foxes, hares and reindeer. The biggest animal is the elk – a deer known as the moose in North America. The elk is a strong swimmer. Its long legs and broad hooves enable it to walk through deep snow. In summer, it spends much of its time in lakes and marshes, feeding on water plants. In winter, it eats bark, pine shoots and twigs.

The elk's main enemies, apart from people, are bears and wolves. Other forest predators include golden eagles, lynxes, owls and wolverines. Fire, often caused by lightning, is another hazard, but it is part of nature's way of renewing the plant life from time to time.

PEOPLE IN THE SUBARCTIC

The taiga, with its long, cold winters, is a thinly populated region. Some of the people who live there are related to the people of the tundra. In northern Europe a group of Sami, sometimes called the 'forest Lapps', spend their lives in villages in the forests. Some rear reindeer, cattle or sheep, while others live by fishing in the rivers and lakes. Many also work in the logging industry. The Siberian taiga is also inhabited by people, such as the Koryak on the Kamchatka peninsula, who raise reindeer, in a similar way to the peoples of the tundra.

The Canadian taiga is the home of several Native American groups, such as the Cree and Chippewa. Before the arrival of Europeans, these people lived by hunting,

fishing and gathering wild plant foods. They did not farm the land because the summers are too short.

Wood, bark and animal skins were used for most everyday items, ranging from household utensils to clothing and houses. For example, some homes consisted of wooden frames covered with animal skins. Others included log cabins and tents called *tepees*. Most of Canada's Native Americans now live in modern homes on reservations. Many of them would like more control over what happens in their own areas.

Forestry is a major activity for people in the taiga. The forests contain softwoods, such as fir and pine, which make good timber because they can be sawn into long, straight planks. Softwoods are also used to make wood pulp and paper. But much of the taiga is remote which makes it costly to transport the wood out of the forest. Minerals are the other main resource of the taiga. Canada's taiga has large deposits of gold, iron ore, lead, zinc and other minerals. The Russian taiga contains diamonds and gold, as well as fuels such as coal and oil.

Left A young Nenet girl standing on a sled in Siberia.

Right top Logging in Ontario, Canada.

Right bottom A Saami reindeer herder and his dog take a rest on the sled during a break in the spring migration in Norway.

Highland climates

FALLING TEMPERATURES

Ecuador is a country on the west coast of South America. It lies across the equator, hence its name. Ecuador means 'equator' in Spanish. Ecuador's coast has a hot climate, but the capital city of Quito lies high in the Andes Mountains, about 2,850 m above sea level. Quito has a cool climate which local people describe as a 'perpetual spring'. This is because temperatures fall by 0.6 °C for every 100 m you climb upwards. The temperature at 2,850 m is about 17 °C cooler than at the coast below.

On steep, high mountains, changes in climate occur over short distances. The tops of high mountains near the equator are often covered with snow and ice

Above *The high Tibetan plateau has a severe climate.*

and they have an icy polar climate, while at the base of the mountains the climate is tropical.

Other climate changes occur when you climb a mountain. It is much windier on the higher slopes than it is lower down. Clouds increase and heavy rain falls as the warm winds blowing up the slopes are chilled. The air also gets thinner and it is often hard to breathe. Also, while daytime temperatures may be high, it may be cold at night.

In Mexico, the people divide their mountainous land into four climate zones. Between sea level and about 900 m is the *tierra caliente* ('hot land'). Between 900 and 1,800 m is the *tierra templada* ('temperate land'). Above 1,800 m is the *tierra fria* ('cold land'). The final zone is the *tierra helada* ('icy land'), which includes the tops of Mexico's tallest peaks.

Because of the effect of the height of the land on climate, scientists regard mountains and high plateau regions, such as Tibet, as a special highland climate zone.

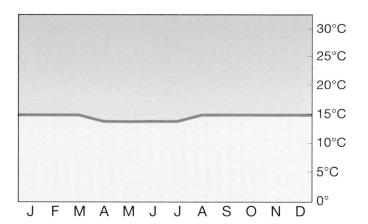

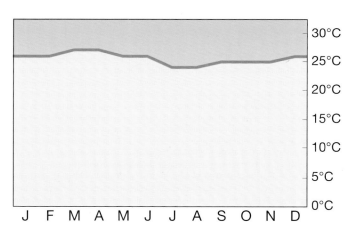

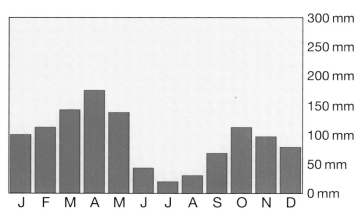

Quito, capital of Ecuador at 2,879 metres above sea level, has a cool climate.

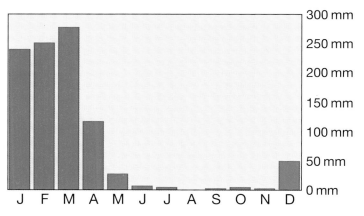

Guayaquil on Ecuador's Pacific coast has a hot climate.

PLANT ZONES

The highest parts of mountains near the equator have an icy polar climate. Some way down the mountain is the snow line. Snow and ice always cover the land above this line, but below it the snow melts for some of the time which enables plants to grow. The snow line around the equator is between about 4,400 and 6,000 m above sea level. In the Alps and Pyrenees in Europe it is lower – between 2,500 and 3,000 m above sea level. Near the poles it is at sea level.

Alpine pastures lie below the snow line. This treeless zone is like the tundra. Here the plants have to withstand low temperatures and fierce winds. Cushion plants are ground-hugging and compact. Their shape gives them protection from the wind and keeps in heat and moisture. Other plants, such as the alpine edelweiss, are covered by white, woolly hairs that protect the plant against the cold. Many plants have long roots which provide good anchors. Because the growing season is usually short, many plants are perennials, growing for several years before they start to bloom.

Near the equator, the high alpine region gets bright sunlight by day,

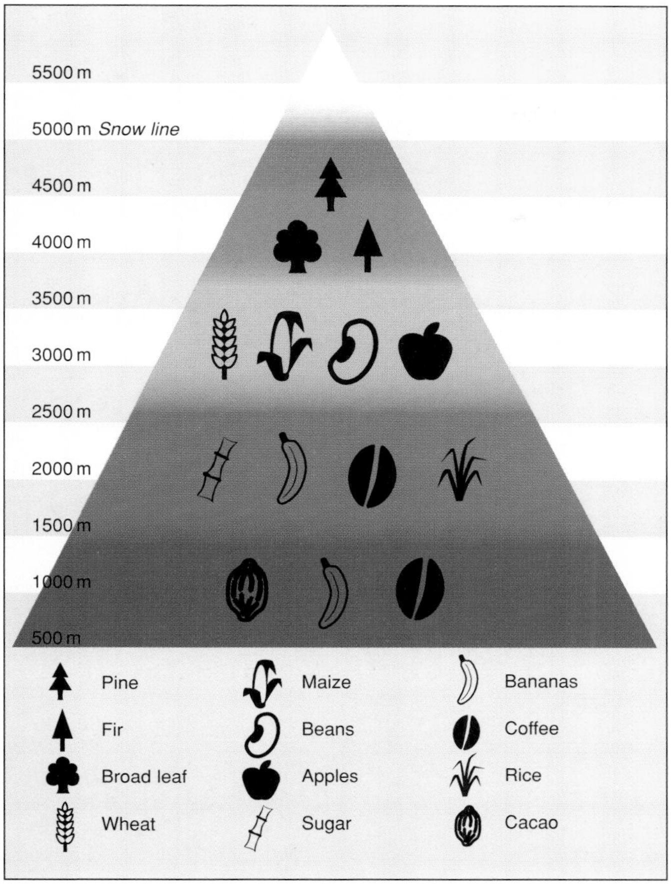
5500 m
5000 m Snow line
4500 m
4000 m
3500 m
3000 m
2500 m
2000 m
1500 m
1000 m
500 m
Pine
Fir
Broad leaf
Wheat
Maize
Beans
Apples
Sugar
Bananas
Coffee
Rice
Cacao

but nights are freezing cold. Some strange plants are adapted to survive this harsh climate, including lobelias, which grow as big as trees here. Giant lobelias produce a fluid like an antifreeze that stops water freezing in its huge rosettes, which close at night.

The alpine zone ends at the tree line, which is similar to the boundary between tundra and subarctic climates. Below the tree line are needleleaf forests, like those in the taiga. Lower down still, the climate becomes temperate and deciduous trees replace the conifers. Eventually, at the bottom of the mountain, is a warm tropical climate zone.

Above *Diagramatic representation of climate zones of a mountain on the equator.*

Left *Giant groundsel in Teleki Valley, Mt Kenya.*

Right *Anazorella – which forms cushions that hug the ground – at 4,500 m in Peru.*

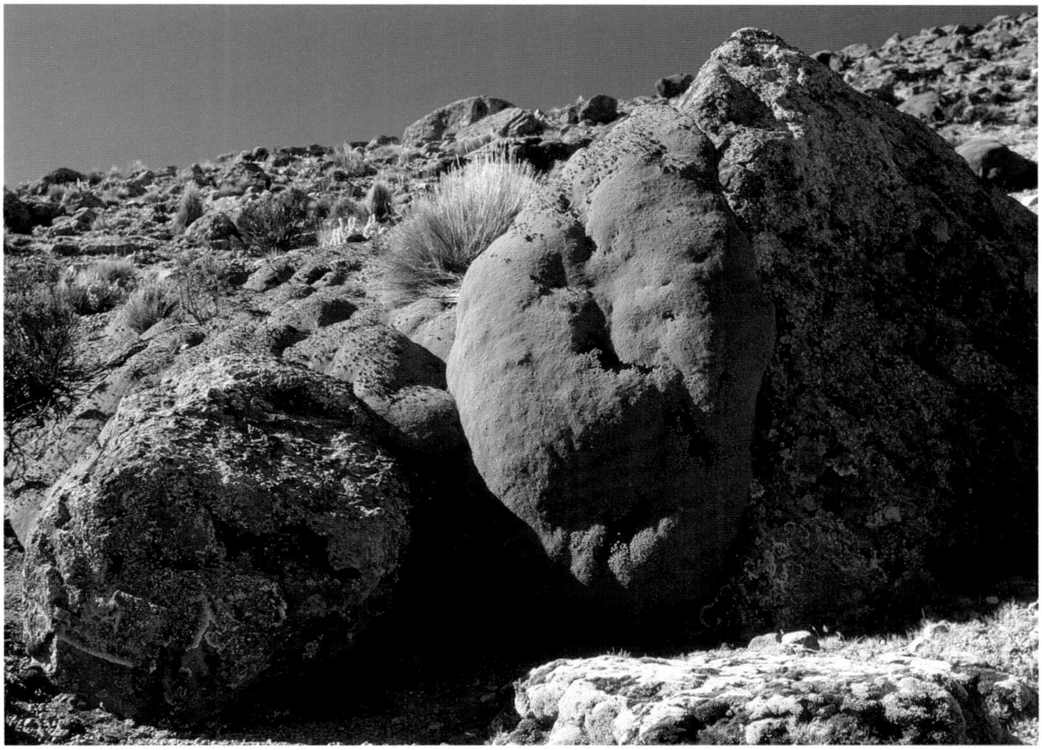

MOUNTAIN ANIMALS

Some mountain animals are migrants, like the caribou and reindeer that move between the taiga and the tundra. They include wild antelopes, goats and sheep. These animals winter in the sheltered forests, but in spring they move up to the alpine pastures to graze. This yearly movement may also be caused by the animals' need to spend some time at lower levels, where the air is thicker and it is easier to breathe.

Mountain animals have special ways of surviving in harsh mountain climates. Some, such as bears, llamas and yaks, have thick coats of fur or hair. Others have layers of fat to protect them. Many small rodents have roundish bodies, short legs and thick fur. Such animals lose less heat from their bodies than tall, slim animals. To survive the winter, some rodents, such as the pika, build up stores of food in summer. Others, such as the marmot (which is a member of the squirrel family), eat their fill in summer, while in winter they hibernate in sheltered areas.

Predators in mountain regions include mountain lions in North America, the snow leopard in Asia

Left Yaks in the Himalayas, India, moving to the spring pastures. Yaks are protected from the cold by their long coats which nearly reach the ground. They can be domesticated and used to carry things.

Right An Alpine marmot in the Italian Alps. Marmots are the largest members of the squirrel family.

and the Siberian tiger. The black serval lives only on the slopes of Mount Kenya. It has a dark coat which absorbs heat from the sun better than the tawny or spotted coats of most servals.

Mountains provide safe nesting sites for birds, but they have to be able to cope with strong winds. Some, such as the mighty condor in the South American Andes Mountains, are large and powerful. This vulture has a wingspan of about 3 m. Other large birds include the lammergeier, which is also a vulture, and the golden eagle. Small birds like the alpine chough stay close to the ground when it is windy.

Above The sure-footed bighorn sheep. Once numerous from southern Canada to northern Mexico, bighorns are now found only in remote mountain areas and national parks in the United States.

Below A lammergeier hovering over rocks.

35

MOUNTAIN PEOPLE

If you visit a mountain area you must wear suitable clothes. People without warm, wind-resistant clothes risk exposure and even death in the harsh mountain climates. Some people suffer from mountain sickness, including headaches, nosebleeds or shortness of breath. Mountain sickness occurs because the air is thin, meaning that there is less oxygen than at sea level. People who want to climb to high peaks usually spend time getting used to conditions at higher and higher levels. This is called acclimatization.

Above Villagers near Cuzco, Peru.

Left A trekking party in East Nepal.

Right A tourist centre in the high Alps, Switzerland.

Mountain sickness does not affect people who live in the mountains, such as the stocky Native Americans whose home is in the Andes. Scientists have found that these people have slightly bigger hearts and lungs than people who live in low-lying areas. They also have more blood cells to carry oxygen quickly around their bodies. Mountain peoples have unusual powers of endurance. Sherpa guides, who lead groups of visitors around the Himalayas (the world's highest mountains) can walk barefoot in the snow without any ill effects.

Many mountain people spend only part of the year at high levels. Farmers who raise livestock often drive their animals up to the alpine pastures in spring, where they spend the summer in small wooden houses. In autumn they return with their animals to their homes at the foot of the mountain. The Kohistani people of Pakistan travel every year between their farms at 600 m above sea level to mountain pasture at 4,300 m. Other herders in central Asia drive their yaks up to levels of more than 5,000 m to find pasture.

Mountains were once regarded as remote and dangerous places. Today's modern transport makes it easier to reach high mountain areas and tourism has become an important activity providing jobs for mountain people. Winter sports, such as skiing, are now especially popular.

Changing climates

ICE AGES

The world's climates have changed many times during the earth's long history. This is shown by the discovery of things in one climate which are normally found in another. For example, scientists have found coal in Antarctica. Coal was formed from plants which grew in warm swamps, hundreds of millions of years ago. The scientists think that, when the coal was formed, Antarctica probably lay much further to the north and that forces inside the earth have moved it to its present position.

The earth has also gone through several great periods of cooling, called ice ages. The most recent is called the Pleistocene Ice Age. It began around 2 million years ago and ended only 10,000 years ago. During this time, vast ice sheets spread over much of the northern hemisphere, covering all of what is now Canada, and spreading south in the United States to Montana, Illinois and New Jersey. In Britain, it reached as far south as London.

It was not cold all the time during the Pleistocene Ice Age. There were long, cold periods when the ice sheets grew in size, but also warm periods, called interglacials, when the ice sheets retreated.

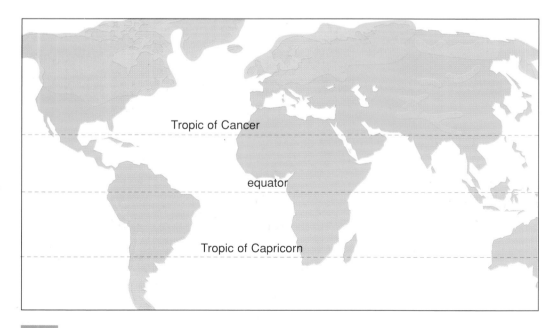

Tropic of Cancer

equator

Tropic of Capricorn

Left *This map shows where the ice spread in the northern hemisphere during the Ice Age. The light-coloured areas are those which were ice-covered.*

Above This valley was formed by the action of glaciers in the Ice Age.

Left and below Fossils of plant and lystrosaurus found in Antarctica.

During some interglacials it was warmer than it is today.

Scientists have put forward theories to explain why ice ages occur. One recent suggestion is that the angle of the earth's tilt varies from time to time. This would move all the earth's climate zones. Scientists think that such changes in the earth's tilt may be linked with other changes in the earth's orbit (path) around the sun. Some scientists think that the Ice Age may not be over and that we may be living in an interglacial period. They suggest that the world may at some time undergo another period of intense cooling. Ice might then cover large areas which now have mild climates.

GLOBAL WARMING

During the Ice Age, so much water was frozen in ice sheets that the sea level fell by about 90 m. Many areas which are now seas were then dry land – North America was joined to Asia, for example, and Britain was joined to France.

Natural forces cause ice ages, but today many scientists think that human actions are changing world climates. In the last 100 years, average world temperatures have risen by between 0.3 and 0.6 °C. In some areas, such as the Antarctic Peninsula, temperatures have risen much faster – by about 2 °C. This has caused the edges of the ice sheets to break up and allow huge icebergs to float away into the sea.

Many scientists think that this 'global warming' is caused by certain 'greenhouse' gases in the lower atmosphere, such as carbon dioxide and methane. Carbon dioxide, which is used by plants, is the most important of these gases. In the last 100 years, the amount of this gas in the atmosphere has greatly increased. This is caused mainly by the burning of coal, oil and natural gas in factories, cars and power plants. Carbon dioxide acts like the glass in a greenhouse – it lets the sun's rays through, but traps some of the heat that is reflected back into the air from the earth's surface. The higher the volume of greenhouse gases in the

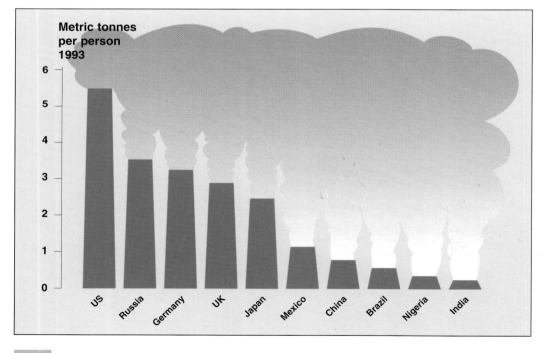

Left *Carbon emissions in 1993.*

Metric tonnes per person 1993

US   Russia   Germany   UK   Japan   Mexico   China   Brazil   Nigeria   India atmosphere, the hotter the earth's climate becomes.

In the last 60 years, global warming has caused a rise in the sea level of about 1.8 mm a year. Some scientists predict further temperature increases of 1.5 to 4.5 °C. If this happens, the world's ice sheets will start to melt and many islands in the oceans will vanish under the waves. Fertile coastal areas, where millions of people now live, will be flooded and storms are likely to increase.

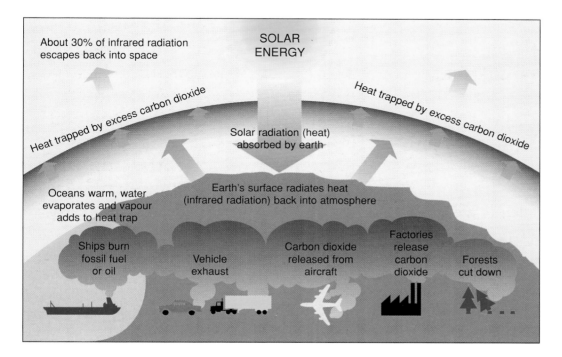

Left *The greenhouse effect.*

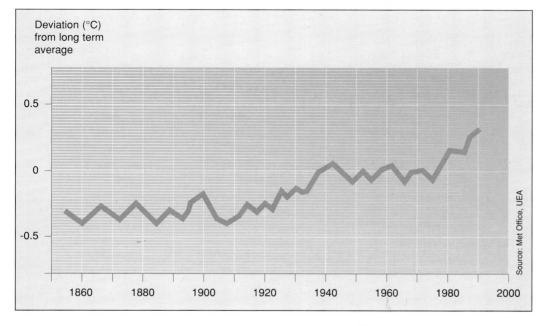

Left *How the average global temperature has risen since 1860.*

POLLUTION AND CLIMATE

In 1982, scientists in Antarctica made an alarming discovery. They found that the ozone layer in the upper atmosphere was much thinner over the icy continent than it had been. Since 1982, this 'ozone hole', which appears in the spring, has got bigger and deeper every year. In 1995, an ozone hole was found over the Arctic.

Ozone is a form of oxygen. Most of it is found in the stratosphere, between about 20 and 40 km above the surface of the earth. Ozone is important because it blocks out most of the sun's harmful ultraviolet radiation which damages plants and animals. It also causes skin cancer and, in the oceans, it kills plankton – the tiny plants and animals that live near the surface, providing food for fish and many other sea creatures.

Scientists believe that ozone holes are caused by chemicals called CFCs. (CFC is short for chlorofluorocarbons.) These chemicals are used in many everyday products, such as aerosol sprays, foams, refrigerators and air-conditioning systems. When these chemicals reach the stratosphere

Left *Icebergs in Antarctica.*

Top right *The ozone layer protects the earth from harmful ultraviolet rays.*

Bottom right *A low coral island. If the sea level rises too much, many low-lying areas, such as this atoll, will disappear under the sea.*

they are broken up by the sun's rays and combine with ozone. In this way, the total amount of ozone in the ozone layer is reduced. Damage to the ozone layer is concentrated around the poles, although traces of it have been detected in other areas.

Scientists have alerted people to the dangers of air pollution, which may cause global warming and an increase in the amount of ultraviolet radiation that reaches the ground. Governments are now working together to reduce carbon dioxide emissions and to replace CFCs with other chemicals that can do the same job. These forms of pollution threaten to change world climates. They also cause serious problems affecting all living things on our planet.

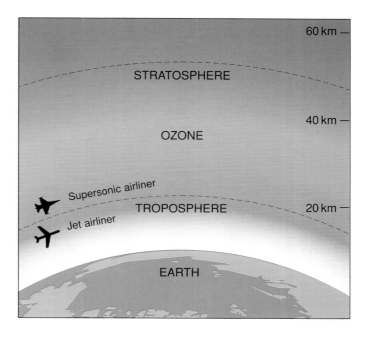

Glossary

Acclimatization Adjusting or adapting to the climate.

Algae Simple plants that live mostly in water or moist soil. They include tiny, single-celled plants and large seaweeds.

Atmosphere The layer of air around the earth.

Camouflage Any means of disguise.

Carbon dioxide A colourless, odourless gas found in the atmosphere. Green plants need carbon dioxide to live and grow. Animals breathe in oxygen and breathe out carbon dioxide. Carbon dioxide is released when coal, oil and natural gas are burned.

Celsius The Celsius scale, named after the Swedish scientist Anders Celsius, is used to measure temperature. The freezing point of water is 0°C. Boiling point is 100°C.

Equator A line of latitude running around the world exactly half way between the North and South Poles.

Evergreen A tree or shrub that remains in leaf throughout the year.

Fungus An organism that cannot make its own food and, instead, absorbs food from its surroundings. Fungi were once classed as part of the plant kingdom. Today scientists classify them separately.

Glacier A body of ice which forms from compressed snow in mountain basins and flows downwards following existing valleys.

Hemisphere Half a sphere.

Hibernation Spending the winter in a resting state.

Ice cap A small ice sheet.

Ice sheet A large body of ice. The earth has two ice sheets – one covers Antarctica and the other Greenland.

Iceberg A floating body of ice in the sea. Only about one-ninth of the ice is visible. The rest is underwater.

Latitude Lines running around the earth parallel to the equator are called lines of latitude. They are measured in degrees between the equator (0 degrees) and the poles (90 degrees North and South). Lines of longitude run around the world at right angles to lines of latitude.

Methane A gas formed when plants decay. It is sometimes called marsh gas.

Migrant A person or animal that moves from one place to another.

Ocean current A flow of water in the ocean. Surface currents are mainly caused by winds. Warm currents heat coastal areas. Cold currents lower temperatures.

Orbit The path followed by a planet, such as the earth, around another planet, such as the sun.

Oxygen A gas which makes up nearly 21 per cent of the atmosphere. Without oxygen we would not be able to breathe.

Ozone layer Protective band in the atmosphere which cuts out most of the sun's harmful utraviolet rays.

Perennial A plant that lives for two or more years or growing seasons.

Pollution The fouling or harming of human, animal or plant life.

Predator An animal that preys upon other animals.

Solar energy Energy that comes from the sun.

Swamp A waterlogged area of ground.

Technology The use of scientific discoveries and inventions to make work easier and life more comfortable.

Temperature The measurement of how hot or cold something is.

Wind chill factor An estimate of the relationship between temperature and wind speed. For example, a strong wind blowing at an air temperature of 6°C feels as though the temperature is −10°C.

Further information

Books

Exploring Weather Catherall, E (Wayland Publishers Ltd, 1990)

People and Climate Hughes, J (Ginn and Company Ltd, 1993)

Polar Lands Chinery, M (Kingfisher Books Ltd, 1992)

How to Survive... at the North Pole Ganeri, A (Simon and Schuster Young Books, 1994)

Conserving Polar Regions James, B (Wayland Publishers Ltd, 1990)

Videos

Climate of Concern, Shell Education Services

From Viewtech films and videos:

Atmospheric Science series

Climate

People Living in other Lands series: *Greenland, Alaska*

Understanding Weather series

Weather Problems and the Environment

Index